Understanding the Elements of the Periodic Table™

THE ALKALINE EARTH METALS

Beryllium, Magnesium, Calcium, Strontium, Barium, Radium

Bridget Heos

rosen publishing's
rosen
central®

New York

For my goddaughters Hannah and Francesca, salt of the earth

Published in 2010 by The Rosen Publishing Group, Inc.
29 East 21st Street, New York, NY 10010

Library of Congress Cataloging-in-Publication Data

Heos, Bridget.
The alkaline earth metals: beryllium, magnesium, calcium, strontium, barium, radium / Bridget Heos.
    p. cm.—(Understanding the elements of the periodic table)
Includes bibliographical references and index.
ISBN-13: 978-1-4358-5331-7 (library binding)
1. Alkaline earth metals—Popular works. 2. Periodic law—Tables—Popular works. I. Title.
QD172.A42H46 2010
546'.39—dc22

                                                    2008055582

Manufactured in the United States of America

On the cover: These diagrams show the number of electrons each alkaline earth metal has. The number varies, but they all have two electrons in their outer shells, making them extremely reactive.

Contents

Introduction   4

Chapter One   Introducing the Alkaline
                  Earth Metals   6

Chapter Two   Atoms, Elements, and
                  Compounds   14

Chapter Three   It's Elemental   21

Chapter Four   Compounds   26

Chapter Five   The Elements and You   32

The Periodic Table of Elements   38

Glossary   40

For More Information   41

For Further Reading   43

Bibliography   44

Index   47

Introduction

Since ancient times, people have known about elements such as gold (Au), silver (Ag), tin (Sn), copper (Cu), lead (Pb), and mercury (Hg). As scientists in the 1600s to 1800s began discovering other elements, they noticed relationships between them. The most accurate and comprehensive description of these relationships, the periodic table, was created by Dmitry Mendeleyev in 1869.

Mendeleyev organized the known elements by their atomic weights (going across, in rows) and by similar properties (going up and down, in columns). However, he determined that based on many of the elements' properties and where they should fit into the periodic table, some of the elements were out of order. Mendeleyev concluded that their accepted atomic weights were probably incorrect. After changing their weights, the table made more sense.

In some cases, he rearranged elements because he thought their atomic weights were incorrect when, in fact, they were correct. Still, he had placed them in the correct order. This is because, as scientists later discovered, the positions of elements are actually determined horizontally not by their atomic weights, but by their atomic numbers—or numbers of protons.

Vertically, the elements in the periodic table are arranged into columns that form families based on similar properties. In this book, you'll learn

While writing a book, Russian chemist Dmitry Mendeleyev (1834–1907) organized the elements according to their relationships to each other. His arrangement became known as the periodic table.

about the alkaline earth metals, a family of highly reactive metals, most of which you've encountered in your daily life as compounds. These elements are found in the second column of the periodic table. Their atoms have two electrons in their outer orbit.

Along with the other elements in the periodic table, the alkaline earth metals are the building blocks for everything in the world, from the bones in our bodies to the sidewalks in our neighborhoods. Here is their story.

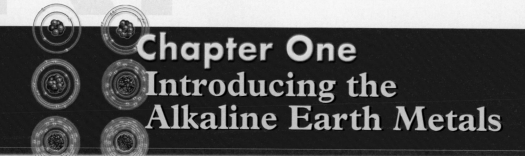

Chapter One
Introducing the
Alkaline Earth Metals

Emeralds. Epsom salt. Milk. Fireworks. Rat poison. Deadly glow-in-the-dark paint. While these things seem to have nothing in common, they all contain elements from the same family: the alkaline earth metals.

Organized in the second column of the periodic table, the group II elements are some of the most reactive in the world. That means they react easily with other elements to form compounds. Therefore, in nature, the alkaline earth metals are always found combined with other elements in compounds, not as pure elements.

To get to know the six alkaline earth metals, let's start by looking at their history. Going down the line, they are:

Beryllium (Be)

The mineral beryl takes on different appearances—many of them beautiful. Emeralds, for instance, are a form of beryl. But so are many purple, blue, brown, yellow, and even peach gemstones. In its pure state, beryl has no color at all. So it's no wonder that people went thousands of years without knowing that all of these stones were the same mineral, beryllium aluminum silicate. Once that was understood, the element beryllium was named after the mineral it helped compose.

Calcium, beryllium, and magnesium, seen here in their elemental form, occur in nature only as compounds.

Frenchman Nicolas-Louis Vauquelin discovered beryllium in 1798. However, it wasn't isolated until 1828. That means that scientists knew beryllium existed as an element before they were able to separate it from a compound.

Magnesium (Mg)

Like beryllium, one of magnesium's compounds was valued long before anybody knew about the element. Epsom salt, later known as magnesium sulfate, was used for healing wounds. Magnesium was discovered in 1755 and named for the district in Greece where it was first found. Englishman Sir Humphry Davy isolated it in 1808.

We now know that magnesium is the eighth most plentiful element in the earth's crust. However, because it is a reactive alkaline earth metal, it is always found in compound form.

Calcium (Ca)

Calcium oxide, or lime, has been used as a plaster since ancient times. In fact, the Great Wall of China and the Roman Colosseum both owe their longevity, in part, to the fact that lime hardens with age. This calcium compound is made by heating limestone or seashells at a high temperature.

You probably think of calcium less in terms of building monuments and more in terms of building bones. That's why they give you milk at school, right? Well, calcium oxide is made by heating limestone at a high

Calcium oxide was used in building the Roman Colosseum—proof that calcium compounds not only build strong teeth but strong monuments, too.

temperature. Limestone, meanwhile, is made of fossilized sea animals. Like you, these creatures need calcium to build strong skeletons (and shells, in their case). To do this, they extract calcium carbonate from seawater. Because the crustaceans' shells and bones make up limestone, so does calcium.

The element, a metal like all the alkaline earths, wasn't isolated until long after structures like the Colosseum were built. The year was 1808. The discoverer was, again, Sir Humphry Davy. Calcium was named for *calx*, which is Latin for "lime."

Strontium (Sr)

Discovered by Adair Crawford and William Cruikshank in 1790, this element is named after a Scottish village, Strontian, where a mineral containing this element was found. In its compound form, strontium is found in the minerals celestite and strontianite. Strontium compounds are used to make red fireworks.

Barium (Ba)

This element is named for the Greek word *barys*, meaning "heavy." In 1774, a scientist named Carl Wilhelm Scheele determined that barium oxide was different from calcium oxide (lime). Humphry Davy then isolated the element in 1808.

Its compounds have many uses. Barium carbonate, for instance, is used as rat poison.

Radium (Ra)

Marie and Pierre Curie discovered radium in 1898 in pitchblende, a uranium ore. They sacrificed riches in the name of science, publishing their methods for extracting radium rather than seeking a patent.

Famous scientist Marie Curie, with her husband, Pierre, discovered radium in 1898. The radioactive element helped many patients suffering from cancer but also made some people sick, including Madame Curie herself.

Radium drugs successfully treated cancer and other diseases. Because of this, people mistook it for a cure-all. Some companies even sold it in water as a drink, dubbed "liquid sunshine." In truth, radium, while helpful in some ways, was also poisonous. Later in this book, you'll read about how the element proved deadly for women painting glow-in-the-dark watches in factories in the 1920s, as well as for Marie Curie.

Meet the Family

Now that you know a little bit about each alkaline earth metal, let's talk about them as a group.

Dmitry Mendeleyev: From Average Student to Father of the Periodic Table

As a boy growing up in Siberia, Dmitry Mendeleyev wasn't the best student. He was good at math and science, but not the classical languages, and that ruined his chances of attending the University of Moscow or the University of St. Petersburg.

However, he was accepted into the St. Petersburg Institute. And when he later reapplied to the University of St. Petersburg for his graduate studies, he was not only admitted but also became well known for his understanding of chemistry.

Mendeleyev created the periodic table while organizing a book, *Principles of Chemistry*. As he decided in which order to write about the elements, he noticed several relationships between them.

He took out sixty-three cards—one for each of the known elements at the time—and wrote on them the element's name, atomic weight, and chemical and physical properties. Then he arranged the cards horizontally according to their atomic weights and vertically according to their properties. He also left space for unknown elements, which he accurately predicted would be discovered eventually. When he finished, he had set up the first periodic table, proving that while he wasn't the best all-round student, he was a scientific genius.

Mendeleyev died in 1907 at age seventy-three. He had been the youngest of seventeen children—maybe that's why he understood elemental families so well.

The group II elements are metals with a silvery color. For the most part, they are abundant in the earth's crust. Calcium is the fifth most abundant element on the earth, for instance and magnesium, the eighth most abundant.

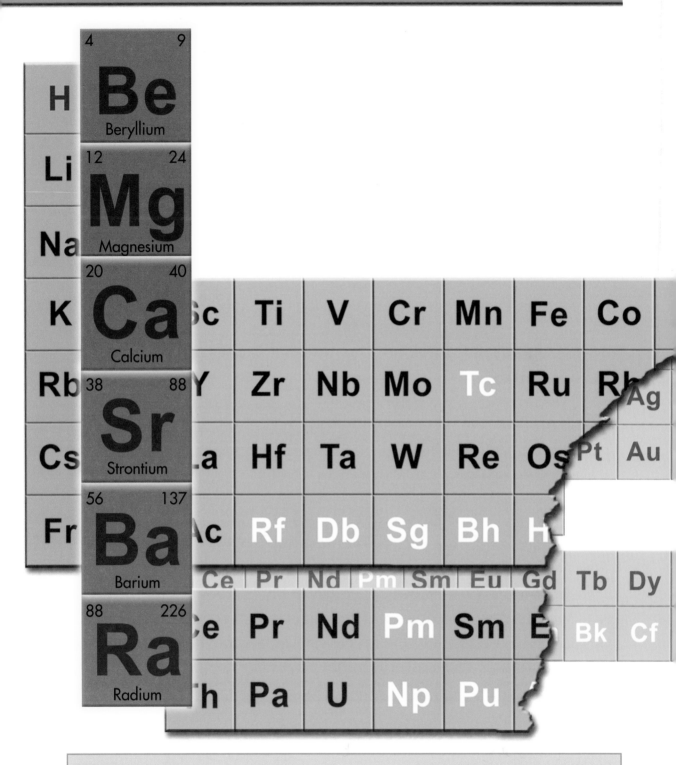

On the periodic table, the alkaline earth metals stand beside the alkali metals, an even more reactive family.

While the alkaline earth metals occur in nature only as compounds, they can be chemically isolated for use in industry. For instance, magnesium as an element is used in cookware, automobile wheels, flares, tracer bullets, and bombs.

As you go down the group, the atomic weight—the average weight of an element's atoms—of the alkaline earth metals increases. Therefore, calcium is heavier than magnesium, and magnesium is heavier than beryllium. The elements also become increasingly reactive.

Meet the Neighbors

The alkaline earth metals are not as reactive as the metals to their left on the periodic table, the alkali metals. Furthermore, the alkali metals have a combining power of one; the alkaline earth metals have a combining power of two. This means that when an alkali metal atom reacts with one other atom, such as chlorine, an alkaline earth atom will react with two of those atoms.

Still, the two groups share some characteristics. Being highly reactive, they are not found as elements in nature but as compounds. Also, the elements in both groups are powerful reducers, meaning they are able to force electrons into other compounds.

The alkaline earth metals are located to the right of the alkali metals in the periodic table because of how the table is arranged. Going horizontally across the table, the elements' atomic numbers increase by increments of one. An atomic number indicates how many protons an atom of an element always has. If the atom didn't have this number of protons, it would not be this element. Take calcium (atomic number 20). It is beside potassium (K) (atomic number 19) to the left and scandium (Sc) (atomic number 21) to the right. If an atom had one less or one more proton than calcium, it would be potassium or scandium.

Chapter Two
Atoms, Elements, and Compounds

Stare at something. Anything. A football, a burrito, your friend. Whatever you're looking at is made up of atoms. All matter is composed of atoms, and everything on the earth is matter. Well, not thoughts and feelings, of course. Love, for instance, is not matter. But a Valentine's Day card is. Anything that has weight or mass is matter and is made up of atoms.

Elements such as the alkaline earth metals have different kinds of atoms. Each element has unique properties, such as the way it reacts with other elements. For instance, through a chemical reaction, atoms of different elements bond, creating a compound that has entirely different properties than the elements.

What gives elements their identity is the number of protons their atoms have. Protons are one of three smaller particles that make up atoms. Protons and neutrons are found inside the atom's nucleus, or center, and electrons orbit the nucleus.

In the alkaline earth metal family, the atoms of each element have two electrons in their outermost level, which is what makes the chemicals so reactive, or quick to join other elements to form compounds. However, the number of total electrons the atoms of an element have varies within the family. Beryllium has four, for instance, while magnesium has twelve, calcium has twenty, strontium has thirty-eight, barium has fifty-six, and

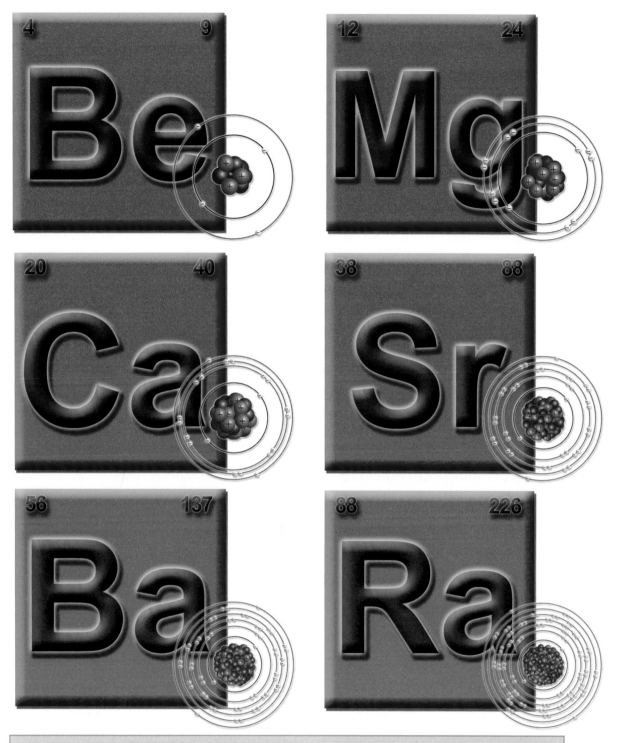

An element's atomic number, shown in the upper-left-hand corner of each tile, refers to the number of protons the element has. Without this number of protons, the element wouldn't be itself. Unless a chemical reaction has occurred, the element has the same number of protons and electrons.

radium has eighty-eight. Look at the periodic table. Do you see these numbers anywhere else? They are the atomic numbers of the elements on the periodic table, which refer to the number of protons. The atoms of the elements have the same number of protons and electrons unless a chemical reaction occurs.

Atoms may lose, gain, or share electrons. Even when they do, the nucleus of each atom remains intact. Therefore, the number of protons remains the same. The only way the protons and neutrons in a nucleus can be removed from the atom is through a nuclear reaction or radioactive decay. On the earth, nuclear reactions are rare.

Chemical reactions occur much more commonly. Here are the four basic types of reactions:

Synthesis occurs when two elements bond, forming a compound. (Element A reacts with element B, forming compound AB. This is represented in symbols by a chemical equation: $A + B \rightarrow AB$.)

In contrast, through **decomposition**, a compound breaks down, becoming isolated elements or simpler compounds. ($AB \rightarrow A + B$.)

In **single displacement**, an element combines with a compound. The element becomes part of a compound, while one of the elements in the compound becomes isolated. ($A + BC \rightarrow AC + B$.)

Double displacement occurs when compounds combine and the elements essentially trade places. ($AB + CD \rightarrow AC + BD$.)

Alkaline earth metals, being so reactive, frequently undergo the chemical reactions described here. For instance, here's a single displacement in which calcium is dropped into water, producing calcium hydroxide and hydrogen gas: $Ca + 2H_2O \rightarrow Ca(OH)_2 + H_2$.

When elements form compounds, chemical bonds hold the atoms together. There are two basic types of bonds:

Barium reacts with water, forming barium hydroxide $(Ba(OH)_2)$. The element is one of the most reactive alkaline earth metals.

In a **covalent bond**, two elements share some of their electrons. The groups of atoms held together by covalent bonds are called molecules.

An **ionic bond** is formed when an atom gives one or more electrons to another atom. The atom that loses electrons becomes a negative ion, and the one that gains electrons becomes a positive ion. The negative and positive ions attract each other, forming an ionic bond.

Almost all alkaline earth metal compounds are ionic. Here's why: Remember how one of their family traits is that each element has two electrons in its outer shell? Well, in an ionic bond, an element with a low number of electrons in the outer shell (such as an alkaline earth metal) reacts with an element with a high number of electrons in the outer shell. The alkaline earth metal gives up its electrons to the other element. Now, the alkaline earth metal is positively charged, and the other element is negatively charged. Because opposite charges attract each other, they are bonded as an ionic compound.

Here is an example of an alkaline earth metal forming an ionic compound: $2Mg + O_2 \rightarrow 2MgO$.

Magnesium atoms give up electrons to oxygen atoms, forming a compound that contains magnesium ions and oxide ions. This compound is magnesium oxide, MgO.

Often, ionic bonds form between metals and nonmetals to form a salt-like substance. This is true of the alkaline earth metals. For instance, Marie Curie used to keep glow-in-the-dark radium salt (probably radium chloride) by her bed as a nightlight, not knowing it was toxic.

Physical Properties

The alkaline earth metals bear a striking family resemblance. In their elemental form, they are metals with a shiny, silvery white color. Once they're exposed to air, however, they form oxides, which dull the shine.

Because the alkaline earth metals have two outer electrons instead of one, they have stronger metallic bonding than the alkali metals. This makes them harder and denser and gives them higher melting points than the alkali metals. (Beryllium has the highest melting point: 2,349 degrees Fahrenheit [1,287 degrees Celsius.]) Some of

A bristlecone pine tree rests on top of dolomite stone in the White Mountains of California. Dolomite is one of many minerals that contain magnesium.

Neutrality Can Be Deadly

While every element has a signature number of protons, the number of neutrons can vary, creating different isotopes of the same element. When an isotope has too many or too few neutrons in its nucleus, the nucleus becomes unstable, or radioactive.

The unstable nuclei release rays of energy that can penetrate the human body. Inside the body, the radioactive particles incite damaging chemical reactions in the body's cells, causing cancer and other diseases.

Radium, plutonium (Pu), and uranium (U) are examples of elements for which all isotopes are radioactive. The other elements in the alkaline earth metal family have some radioactive isotopes but some stable ones, too.

the alkaline earths are also known for their vibrancy when they burn. Magnesium burns brilliant white, calcium burns yellow-red, strontium and radium burn crimson, and barium burns apple green.

Compounds of alkaline earth metals are very common. Magnesium, for instance, is found in rocks in the minerals carnallite, magnesite, and dolomite, and calcium is found in chalk, limestone, gypsum, and anhydrite.

Chemical Properties

You learned earlier in the chapter how alkaline earth metals form ionic bonds with other elements. They lose their outermost electrons to the other element, and the two are then bonded through their positive and negative charges. The property of easily losing electrons is called having a strong reducing power. Going down the column in the periodic table, the family's reducing power—and reactivity—becomes stronger. Thus, beryllium

All alkaline earth metals react to dilute acids. In the test tubes seen here on the left, calcium and magnesium react vigorously to hydrochloric acid. To the right, elements from other families react differently—zinc, slowly, and copper, not at all.

has the least reducing power of the metals while barium is so powerful a reducer that it has to be stored under oil in the lab. Radium is the most powerful reducer.

There are some chemical properties that all of the alkaline earth elements share. They all react with the elements oxygen (O) and chlorine (Cl), and when added to dilute acids, they all react through the release of hydrogen gas.

Like every family, the elements have unique characteristics, too. Individually, calcium reacts rapidly with water, forming calcium hydroxide and hydrogen gas. Magnesium reacts with water, too, but the water has to be boiling. And beryllium doesn't react to water or steam—not even when the metal is red hot.

Radium is unique in the family because it is the only one that is radioactive. This means that the nuclei of its atoms, unlike most nuclei, aren't stable. Instead, they disintegrate, releasing energy, in radium's case, in the form of alpha particles and gamma rays. For this reason, even being near radium can cause cancer or other disorders. And you definitely wouldn't want to eat radium. Unfortunately, when radium was first discovered, it was considered healthy at best, harmless at worst, and people did indeed consume it.

Chapter Three
It's Elemental

While alkaline earth metals aren't found in their elemental form in nature, scientists can isolate the elements through chemical reactions.

Beryllium

In the case of beryllium, scientists reduce beryllium fluoride with magnesium metal. Once isolated, beryllium has many uses due to its lightness, high melting point, and thermal conductivity. Scientists alloy (mix) it with copper or nickel (Ni) to create a stronger metal for springs, electrical contacts, and non-sparking tools. Beryllium is also used to make airplanes, spacecraft, missiles, X-ray windows, and computer parts.

Scientists have to be careful when working with beryllium. Both the element and its compounds are toxic. In years past, researchers didn't know this. To verify that an element was beryllium, they would taste it for sweetness. In fact, beryllium was originally called glucinium, from the Greek word for "sweet," because of its sweet flavor.

Dust from the metal is also toxic. It causes lung damage called berylliosis. Unfortunately, beryllium gets into our environment through industrial smoke.

Magnesium

Magnesium is an essential nutrient. In the body, it is found as a cation— a positively charged ion, Mg^{+2}. Magnesium forms this ion because, as you recall, alkaline earth metals form ionic bonds by giving up two of their electrons, thus becoming positively charged. You can get your daily intake of magnesium by eating almonds and other nuts, green vegetables, peas and beans, or by drinking tap water if your town has hard water.

Sea water is the main source of the 748,000 metric tons of magnesium produced annually worldwide. As such, magnesium is the most widely produced alkaline earth metal. Isolated through electrolysis of fused magnesium chloride, the element is used in making flares, fireworks, and bombs, and in many other applications.

Magnesium has a dangerous side, however. It catches fire easily. When heated to a high temperature in air, it ignites. Burning at 3,600°F (1,982°C), the fire is extremely hot. (In comparison, paper burns at 454°F [234°C]).

Even worse, a magnesium fire is hard to put out. Water and carbon dioxide fire extinguishers make things worse. Hot magnesium reacts with both, spreading the fire or causing explosions. Instead, the fire has to be smothered with sand, but it is difficult to get close enough to the flame to do that.

What do these foods have in common? They all contain magnesium. Children ages nine to thirteen require 240 mg of this important nutrient each day. Older teens need 360 to 410 mg.

Calcium

Calcium is the fifth most abundant element on the earth, and calcium and its compounds are primarily derived from limestone.

Scientists isolate the element through electrolysis of fused calcium chloride and calcium fluoride. Calcium metal is used to make alloys and as a reducer to extract other metals, such as thorium (Th) and uranium, from their compounds.

Strontium

Like calcium and magnesium, strontium is isolated through electrolysis. The element isn't used a great deal, except to produce color television picture tubes, in ceramics and glass manufacturing, and sometimes as an alloy.

Still, strontium is widespread in the environment. One of strontium's sixteen unstable isotopes, strontium-90, is found in the soil because it is in the fallout from nuclear weapon testing done in the 1950s and 1960s, and from the 1986 Chernobyl nuclear power plant accident in the Ukraine. Because strontium is similar to calcium,

This little boy lives in an apartment near the former Chernobyl nuclear plant. The machine beside him, a Geiger counter, shows that more than 250 times the normal amount of radiation exists in the boy's front yard—the result of nuclear fallout from an accident at the plant.

strontium-90 is taken up by plants in place of calcium. This isotope gets into the food chain, and when a person consumes it, the body mistakes it for calcium and stores it in the skeleton, where it can cause bone tumors.

While dangerous in some ways, strontium isn't intrinsically bad. Stable forms of it are not particularly harmful. Radioactive strontium-85 is used in medicine. Even strontium-90 has been used in the treatment of eye disease and bone cancer.

Barium

Found in nature as a sulfate and carbonate, barium is isolated through electrolysis of barium chloride. As an element, barium is used mainly in vacuum tubes. It is very reactive and has to be stored under oxygen-free liquids to prevent oxidation.

Radium

While radium is rarely used today, its discovery revolutionized the way scientists viewed the atom.

Marie Curie was a graduate physics student in the late 1800s when she began studying uranium ore. She was measuring the radioactive rays given off by uranium and thorium in the ore when she made a discovery: the ore's radioactivity was greater than it should be for uranium and thorium alone. She hypothesized that the ore contained another unknown element.

With husband and fellow physicist Pierre Curie, she set out to isolate the element. They thought it would take weeks or months. Instead, it took years. They discovered multiple elements within uranium ore, including radium.

Together, they produced radium chloride from the ore. After Pierre Curie's death, Marie Curie isolated pure radium. She did this through the electrolysis of radium chloride with a mercury cathode, and distilling it in an atmosphere of hydrogen.

Sir Humphry Davy: Discoverer of Four Alkaline Earth Metals

Born in southwest England in 1778, Humphry Davy didn't like school when he was a boy. He chose instead to explore things that interested him on his own. He was interested in poetry and writing stories. Then, when he was sixteen years old, his father died. His family, which had five children, was left in debt. Davy dropped out of school.

At seventeen, he became an apprentice to a pharmacist. His job was to mix potions. At twenty, he got a job at Dr. Thomas Beddoes's Pneumatic Institution, where he discovered that N_2O, laughing gas, could be used as anesthesia in surgery.

He became famous for that and was invited to teach at the Royal Institution of Great Britain, a prestigious research institute. While there, through electrolysis, he isolated sodium and potassium for the first time. In 1808, also using electrolysis, he isolated four of the six alkaline earth metals: barium, calcium, magnesium, and strontium.

Curie learned that radium, as a radioactive element, behaves differently from stable elements. As radium releases radioactive particles, its atoms disintegrate. Eventually, the atoms break down to become helium (He) and lead. Previously, scientists believed atoms had to be stable. They now knew that atoms could spontaneously change.

Later, Curie was excited by the healing potential of radium rays. By the 1920s, radium had been proven to treat cancer. It was used in radiation therapy for many years.

Unfortunately, people didn't realize at the time that while radium rays could kill cancerous cells, they could also kill healthy ones. In 1934, Madame Curie died at age sixty-seven of leukemia—the result of working with radium.

25

Chapter Four
Compounds

When two or more elements bond to form compounds, they become something other than the sum of their parts. The alkaline earth metals, for instance, are silvery white metals. Their compounds, however, can be anything from expensive emeralds to over-the-counter antacids. What accounts for this change in identity? The elements haven't simply mixed together. Rather, they've undergone a chemical reaction.

Mixtures are like chocolate and vanilla swirl ice cream, which, predictably, tastes like chocolate and vanilla ice cream. You probably wouldn't even mind eating them separately. In contrast, compounds are like bread, which does, in fact, undergo chemical changes when a mixture of yeast, flour, salt, and water is baked into bread. You might not want to eat the ingredients separately, but as baked bread, they taste delicious.

When elements bond to form a compound, the compound is something entirely different from the elements. Its physical makeup is different, for one thing. For instance, the element calcium is a metal; the compound calcium carbonate is a rock. Their freezing and melting points are different, they react differently to other elements and compounds, and they have different uses. In short, the compound's properties are nothing like the elements' properties.

This oddly shaped limestone, which scientists call tufa, is the result of calcium from an underwater spring mixing with carbonate-rich lake water. After the towers formed underwater, the lake level dropped.

Unlike mixtures, compounds always contain elements in the same proportion. For instance, water, H_2O, always contains one oxygen atom for every two hydrogen atoms. Also unlike mixtures, elements in a compound can be separated only by a chemical reaction.

In formulas and names for ionic compounds, the positively charged ion is listed first, so you'll find the alkaline earth metals at the beginning of many compound names.

Beryllium

Beryllium is most well known for its minerals. Minerals are solids composed of a single compound—and in a few cases, a single element—that has a crystalline form. The most common beryllium minerals are beryl and bertrandite. The chemical formula for beryl is $Be_3(Al_2(SiO_3))_6$.

When beryl contains traces of chromium (Cr), it becomes an emerald. Only deep green beryl is considered an emerald. Pale green beryl is simply called "green beryl." Oddly, chromium causes the opposite effect in the mineral corundum. It makes it a red ruby. When beryl contains traces of iron (Fe), it becomes blue (aquamarine). The reason elements change the color of minerals is that the atoms absorb portions of the light spectrum.

Beryl, a mineral containing beryllium, is best known for its emerald form. The deep green hue occurs when beryl contains traces of chromium.

The colors the atoms do not absorb are reflected. These are the colors we see. Iron in beryl does not absorb blue, for instance, so that is the color we see in aquamarine.

Magnesium

Magnesium is perhaps best known for its role in life-sustaining and healing compounds, which are also extremely stable. In the human body, magnesium ions are important for healthy bones, teeth, muscles, and nerves, and in plants, magnesium is essential to growth. Chlorophyll is what allows plants to trap sunlight, the energy needed to convert carbon dioxide and water into glucose and oxygen. Though not the most plentiful element in chlorophyll, magnesium ions are at the heart of chlorophyll molecules. The chemical formula for chlorophyll is: $C_{55}H_{72}MgN_4O_5$.

Many magnesium compounds are used for medicinal purposes, including magnesium hydroxide, chloride, citrate, and sulfate. Soaking in magnesium sulfate, commonly known as Epsom salt ($MgSO_4$), soothes muscles and relaxes the body, and magnesium hydroxide in water, known as milk of magnesia ($Mg(OH)_2$), is an antacid and laxative.

Not all magnesium compounds are healthy, however. Magnesium is a component of asbestos, a naturally occurring mineral that, when its dust is inhaled over long periods of time, can cause lung cancer and other health problems.

Bitter Water, Healing Salt

You can lead a cow to water, but you can't make it drink. Not when it tastes like Epsom salt, at least. That's what farmer Henry Wicker in Epsom, England, learned in 1618. During a drought, he brought his thirsty cattle to the town commons to drink from a watering hole. Surprisingly, they refused to drink it. He tasted the water himself and understood why: it tasted terrible. The bitterness was soon discovered to be caused by Epsom salt, or magnesium sulfate. While it wasn't tasty to drink, people found that soaking in Epsom salt water healed sores. Today, doctors still recommend Epsom salt for its healing properties.

Calcium

Calcium compounds build both strong skeletons and strong buildings. In the human body, calcium hydroxyapatite ($Ca_{10}(PO_4)_6(OH)_2$) is the primary mineral in bones and teeth. Since ancient times, limestone ($CaCO_3$) has been used in building blocks and has been heated to create lime (CaO) for use in mortar or cement.

Today, lime has many other uses, including pollution control and removing unwanted sand when transforming iron ore into steel. In fact, it is usually one of the top five chemicals produced in the United States. Other calcium compounds are used in everything from deodorant to ice cream.

Strontium

Strontium is found naturally in celestite, a mineral named for its sky blue or "celestial" color, and in strontianite, a mineral with a snowy color. Strontium titanate, a man-made gemstone marketed in decades past with names such as Marvelite and Fabulite, was once a substitute for diamonds. Because it was easily scratched, other fake diamonds have since become more popular. Today, strontium compounds are mainly used in making colored fireworks and flares, as well as in glass and ceramics manufacturing. They are also used in producing ferrite magnets and in refining zinc.

Barium

Barium compounds have diverse uses. Barium sulfate is used for paint, glassmaking, and body imaging with X-rays. Barite is used to make rubber. Barium carbonate is used as a rat poison. Barium nitrate and chlorate create green flares and fireworks.

However, the most common use of barium compounds is in the oil industry. In the old days, oil drillers sometimes encountered "gushers." A gusher occurred when oil rushed up the hole and sprayed everywhere, wasting the oil and causing a flammable mess. Now, barite is added to the petroleum as the hole is drilled. It mixes with the oil and weighs it down so that it doesn't gush out of the hole.

In nature, barium is found in the minerals barite and witherite.

Radium

As you learned in chapter 3, in the 1920s, radium was discovered to be a successful cancer treatment. People mistook this as a green light to use radium freely. They believed that at best it was a cure-all and at worst it was harmless.

People used radium compounds in commercial products, ignorant, at first, of its effects. In the 1920s, Radium Dial Co. in Illinois and United States Radium Co. in New Jersey made glow-in-the-dark watches. To do this, they hired young women to paint the numbers with a mixture that

So-called radium girls were employed to paint glow-in-the-dark watches. Because the paint contained radium, many workers suffered severe health problems. Some of the women sued their employer and received a settlement. Sadly, all of the plaintiffs died at a young age.

contained radium sulfate. The women were encouraged to wet and shape the paintbrush by twirling it in their mouths.

At first, the young workers thought the job was fun. They even painted their nails and teeth with the radium paint so that they would glow in the dark. But soon, they suffered terrible toothaches, which foreshadowed much worse health problems: tumors, blood disease, and bone fractures. Many of the workers eventually died from radium poisoning, and it was revealed that company officials knew the dangers of radium.

Radium has taught scientists many lessons. For the public, it has offered a heartbreaking example of what happens when companies use dangerous or unknown chemicals without concern for their workers.

Chapter Five
The Elements and You

As you've read many times by now, the alkaline earth metals, in nature, are found as compounds. The metals are different, in that way, from gold and silver and other standalone elements. For that reason, you'll probably never see them in their elemental form, unless, of course, you become a scientist who works with the metals.

Here are some places you might see their compounds, however.

Beryllium

While you're probably not in the market for a beryl gemstone, you can see them in books about minerals and precious stones, available at your library or at museums. If you're adventurous, you can look for beryl yourself by becoming a rock hound—a person who searches for rocks and minerals after researching where they're likely to be found.

Amateur rock hound Steve Brancato was searching Mount Antero in central Colorado when he found a 37-by-25-inch (94-by-64-centimeter) rock that contained one hundred aquamarine crystals—the largest known specimen in North America. He named it Diane's Pocket after his mother and donated it to the Denver Museum of Nature and Science.

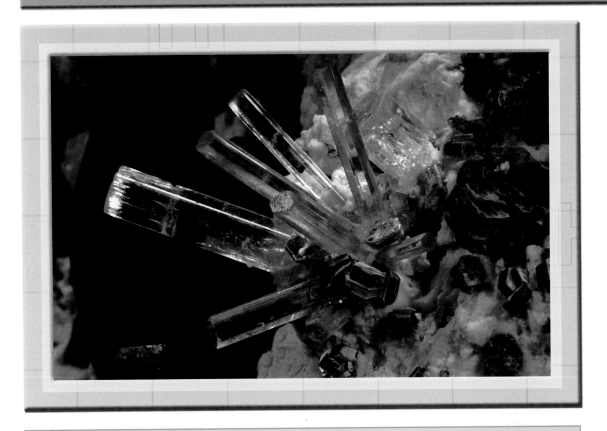

Diane's Pocket, now housed in the Denver Museum of Nature and Science, is the largest known aquamarine specimen in North America. An amateur rock hound found it in central Colorado.

Magnesium

If you eat a balanced diet, you encounter magnesium every day. It is found in green vegetables (because of its role in chlorophyll molecules), nuts, seeds, and beans, and many other foods—even chocolate! After you eat magnesium-rich foods, the nutrient is absorbed through the small intestines and used for more than three hundred biochemical reactions in the body.

A magnesium compound you might want to use if you dance or play sports is Epsom salt. Soaking in Epsom salt water relieves muscle pain. Magnesium is the eighth most plentiful element in the earth's crust, and rivers erode magnesium compounds from the land and carry them to the sea. In the sea, these compounds are converted to magnesium chloride. If you don't live by the ocean, you can soak in sea bath salts instead, which contain magnesium chloride and other minerals.

Calcium

You probably use at least one calcium compound every day. They're everywhere. Calcium is in dairy products because the ion plays a role in the protein molecules that make up milk. It is also found in green vegetables.

Seashells, egg shells, and rocks such as limestone, chalk, and marble contain calcium carbonate. Several foods, including ice cream, have calcium alginate as a thickener. Calcium chloride is used to remove ice from roads. Calcium hypochlorite is a swimming pool disinfectant and deodorant ingredient. Calcium stearate is in cement. And calcium tungstate is in fluorescent lights.

While calcium is all around you, unfortunately, you probably don't have enough of it in your bones. Of boys ages twelve to nineteen, only 36 percent get the calcium they need. For girls the same age, only about 13 percent get enough calcium.

Milk is perhaps the best-known substance containing calcium. However, calcium compounds are all around you—from the cement beneath your feet to the fluorescent lights overhead.

Waiter, There's an Element in My Soup

Several elements, including calcium and magnesium, are important to your health. Here are some of them:

Calcium builds strong bones and teeth. Boys and girls ages nine to eighteen need 1,300 milligrams per day. Sources include dairy products, tofu, and many green vegetables.

Iron is important to cell growth and the delivery of oxygen to cells. In the body, iron is found in the hemoglobin in red blood cells. Meat, fish, and poultry are good sources of iron because animals also store iron in their hemoglobin. The daily requirement is 8 mg for boys and girls nine to twelve years old, 11 mg for boys fourteen to eighteen, and for girls the same age, 15 mg.

Magnesium is good for the heart, immune system, muscles, nerves, and bones. For a nine- to thirteen-year-old boy or girl, 240 mg of magnesium are required each day. For fourteen- to eighteen-year-olds, that amount jumps to 410 mg for boys and 360 mg for girls. Magnesium is found in a variety of foods. By eating a balanced diet, most people get their daily requirement.

Potassium helps cells, tissues, and muscles function properly. It is in meat, soy products, and several fruits and vegetables. Children ages nine to eighteen need 4.5–4.7 grams per day.

Zinc is essential for healthy cells, a strong immune system, and for children, normal growth. Children ages nine to thirteen require 8 mg daily, and fourteen- to eighteen-year-olds need 9–11 milligrams. Sources include meat, poultry, shellfish, beans, and nuts.

Strontium

Other than the radioactive strontium-90 found around the world because of nuclear tests and accidents, the element is pretty benign. One of its compounds, strontium chromate, is extremely poisonous, but that's mainly because of the chromium.

Besides that, strontium occurs naturally in rocks and soil, and in food, water, and the air. Strontium carbonate is used in making the special glass in color television tubes, and on the Fourth of July, you can see strontium compounds in red fireworks.

Barium

If you're ever having unexplained pain in your abdomen, doctors might use barium sulfate to find out what's wrong. You would swallow the chalky substance mixed with water, and the mixture would coat your esophagus, stomach, and small intestines. Doctors would take an X-ray, and the barium sulfate would block the X-rays, making your organs appear on the X-ray as bone would. Remember, barium compounds are poisonous. But because

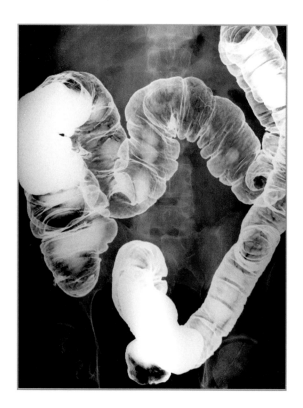

Here, barium is used to view a large intestine through an X-ray. The barium compound makes the organ visible, just as bone is visible on an X-ray.

barium sulfate isn't water soluble, the body doesn't absorb it during the test.

Barium compounds occur naturally in rocks and soil, and are used in green fireworks (barium burns green) and rat poison.

Radium

Of course, you wouldn't want to handle radium, and it's unlikely that you ever will. However, you do have some danger of encountering radon, the gas radium produces when it undergoes nuclear decay.

This is because uranium is a common element in soil. When uranium breaks down by nuclear decay, it produces radium, which is dangerous but, as a solid, remains in the soil. Radium then decays and produces radon. Because radon is a gas, it seeps out of the soil and into homes through cracks in the foundation. It mixes with air, and people can inhale it when they breathe. When radon decays, it forms other radioactive isotopes that are solids. These solids get trapped in people's lungs, where the radiation that they produce can cause lung cancer—even if the person has never smoked. The Environmental Protection Agency estimates that radon kills twenty-one thousand people a year in the United States. After smoking, it is the second-leading cause of lung cancer in the United States.

While you can't see or smell radon, you can buy a radon test kit at the hardware store. It will tell you if your home has elevated levels of this dangerous gas.

The Periodic Table of Elements

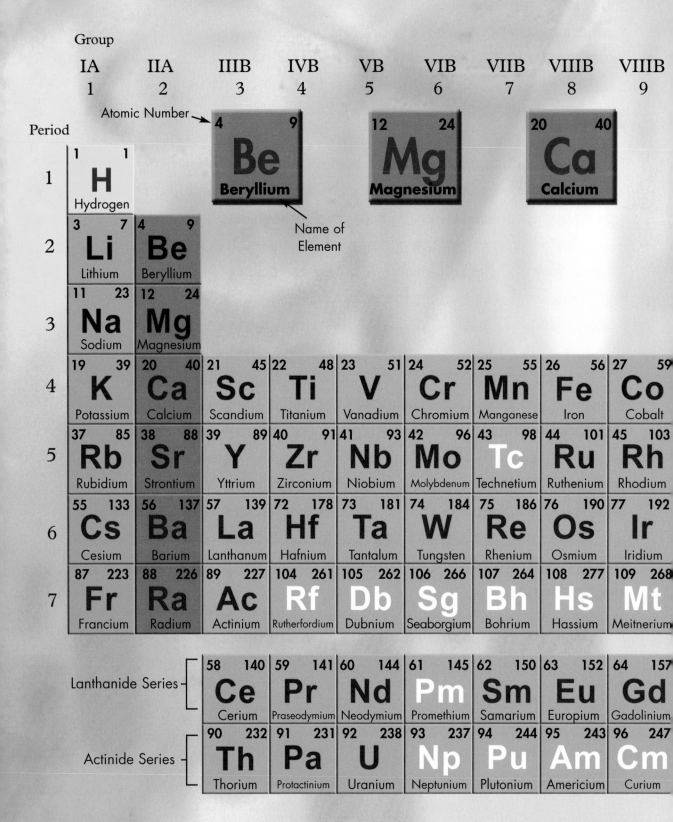

Group

IA	IIA	IIIB	IVB	VB	VIB	VIIB	VIIIB	VIIIB
1	2	3	4	5	6	7	8	9

Atomic Number

4	9
Be	
Beryllium	

Name of Element

12	24
Mg	
Magnesium	

20	40
Ca	
Calcium	

Period

1 — 1 | 1 — **H** — Hydrogen

2 — 3 | 7 — **Li** — Lithium ; 4 | 9 — **Be** — Beryllium

3 — 11 | 23 — **Na** — Sodium ; 12 | 24 — **Mg** — Magnesium

4 — 19 | 39 — **K** — Potassium ; 20 | 40 — **Ca** — Calcium ; 21 | 45 — **Sc** — Scandium ; 22 | 48 — **Ti** — Titanium ; 23 | 51 — **V** — Vanadium ; 24 | 52 — **Cr** — Chromium ; 25 | 55 — **Mn** — Manganese ; 26 | 56 — **Fe** — Iron ; 27 | 59 — **Co** — Cobalt

5 — 37 | 85 — **Rb** — Rubidium ; 38 | 88 — **Sr** — Strontium ; 39 | 89 — **Y** — Yttrium ; 40 | 91 — **Zr** — Zirconium ; 41 | 93 — **Nb** — Niobium ; 42 | 96 — **Mo** — Molybdenum ; 43 | 98 — **Tc** — Technetium ; 44 | 101 — **Ru** — Ruthenium ; 45 | 103 — **Rh** — Rhodium

6 — 55 | 133 — **Cs** — Cesium ; 56 | 137 — **Ba** — Barium ; 57 | 139 — **La** — Lanthanum ; 72 | 178 — **Hf** — Hafnium ; 73 | 181 — **Ta** — Tantalum ; 74 | 184 — **W** — Tungsten ; 75 | 186 — **Re** — Rhenium ; 76 | 190 — **Os** — Osmium ; 77 | 192 — **Ir** — Iridium

7 — 87 | 223 — **Fr** — Francium ; 88 | 226 — **Ra** — Radium ; 89 | 227 — **Ac** — Actinium ; 104 | 261 — **Rf** — Rutherfordium ; 105 | 262 — **Db** — Dubnium ; 106 | 266 — **Sg** — Seaborgium ; 107 | 264 — **Bh** — Bohrium ; 108 | 277 — **Hs** — Hassium ; 109 | 268 — **Mt** — Meitnerium

Lanthanide Series —

58	140	59	141	60	144	61	145	62	150	63	152	64	157
Ce		**Pr**		**Nd**		**Pm**		**Sm**		**Eu**		**Gd**	
Cerium		Praseodymium		Neodymium		Promethium		Samarium		Europium		Gadolinium	

Actinide Series —

90	232	91	231	92	238	93	237	94	244	95	243	96	247
Th		**Pa**		**U**		**Np**		**Pu**		**Am**		**Cm**	
Thorium		Protactinium		Uranium		Neptunium		Plutonium		Americium		Curium	

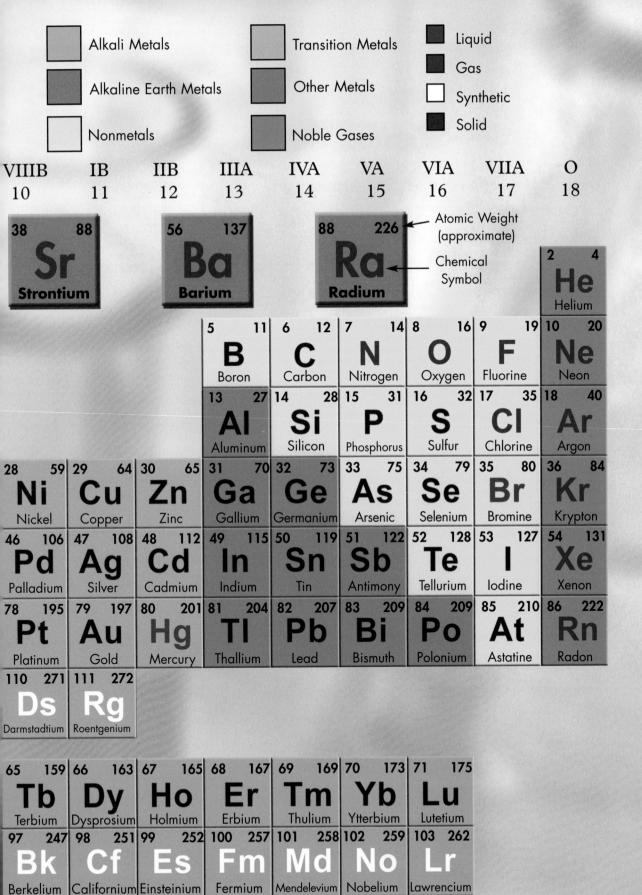

Glossary

alkaline earth metals The elements in group II of the periodic table, characterized by having atoms with two electrons in their outer shell, making them highly reactive.

atom The smallest particle of an element that retains the properties of the element; a particle having a nucleus that contains protons and neutrons, and electrons orbiting the nucleus.

compound A substance that contains two or more elements, always in the same proportion.

covalent bond The connection between two atoms formed when they share electrons.

electron Negatively charged particle that orbits the nucleus of an atom.

element A substance in which all atoms have the same number of protons.

ionic bond The attraction between negatively and positively charged ions that forms when one atom takes electrons from another atom.

isotope A variation of an element that contains different numbers of neutrons.

mineral Matter found in the earth that contains only one element or one compound in a crystalline form.

molecule The smallest particle of an element or covalent compound retaining the characteristics of the element or compound.

neutron Uncharged particle that exists in the nucleus of an atom.

nucleus The center of an atom, containing neutrons and protons.

proton Positively charged particle that exists in the nucleus of an atom.

radioactive The property of an atom that causes it to spontaneously disintegrate into a new element while emitting particles and energy rays.

reactivity An substance's tendency to interact with other substances.

For More Information

American Chemical Society
1155 Sixteenth Street NW
Washington, DC 20036
Web site: http://www.acs.org
The congressionally chartered organization represents professionals in
all areas of chemistry.

International Union of Pure and Applied Chemistry
IUPAC Secretariat
P.O. Box 13757
Research Triangle Park, NC 27709-3757
Web site: http://www.iupac.org
This international union of national chemical organizations promotes the
advancement of chemistry for the good of mankind.

National Academy of Sciences
500 Fifth Street NW
Washington, DC 20001
Web site: http://www.nas.edu/usnc-iupac
This institute is charged with advising the nation on matters related to
science, engineering, and medicine.

Royal Chemistry Society
Burlington House
Piccadilly, London W1J 0BA
England
Web site: http://www.rsc.org

This is the largest organization in Europe dedicated to advancing chemistry studies.

Science National Honor Society
Board of Directors
2929 Fountainview
Houston, TX 77057
Web site: http://www.sciencenhs.org
This national organization is composed of local chapters of high school students interested in science.

U.S. Department of Energy National Science Bowl
Office of Science (SC-27)
U.S. Department of Energy
1000 Independence Avenue SW
Washington, DC 20585
(202) 586-7231
Web site: http://www.scied.science.doe.gov/nsb
This is a competition for high school science students.

Web Sites

Due to the changing nature of Internet links, Rosen Publishing has developed an online list of Web sites related to the subject of this book. This site is updated regularly. Please use this link to access the list:

http://www.rosenlinks.com/uept/taem

For Further Reading

Brown, Cynthia Light. *Amazing Kitchen Chemistry Projects You Can Make Yourself* (Build It Yourself). White River Junction, VT: Nomad Press, 2008.

Clark, Claudia. *Radium Girls: Women and Industrial Health Reform, 1910–1935*. Chapel Hill, NC: The University of North Carolina Press, 1997.

Cooper, Chris. *Matter* (DK Eyewitness Books). New York, NY: DK Children, 1999.

Emsley, John. *Nature's Building Blocks: An A–Z Guide to the Elements*. Oxford, England: Oxford University Press, 2003.

Emsley, John. *Vanity, Vitality, Virility: The Science Behind the Products You Love to Buy*. Oxford, England: Oxford University Press, 2006.

Ham, Becky. *The Periodic Table* (Essential Chemistry). New York, NY: Chelsea House Publications, 2008.

Hill, Lisa. *The Properties of Elements and Compounds* (Sci-Hi: Physical Science). Chicago, IL: Heinemann Raintree, 2008.

McDowell, Julie. *Metals* (Essential Chemistry). New York, NY: Chelsea House Publications, 2008.

Newmark, Ann. *Chemistry* (DK Eyewitness Books). New York, NY: DK Children, 2005.

Saunders, Nigel. *Exploring Atoms and Molecules* (Exploring Physical Science). New York, NY: Rosen Publishing, 2007.

Bibliography

Agency for Toxic Substances and Disease Registry. "Public Health Statement for Strontium." Retrieved December 3, 2008 (http://www.atsdr.cdc.gov/toxprofiles/phs159.html).

"Alkaline Earth Metals: Magnesium." Retrieved December 2, 2008 (http://science.jrank.org/pages/220/Alkaline-Earth-Metals-Magnesium.html).

American Chemistry Council. "Awe-Inspiring Fireworks." Retrieved September 12, 2008 (http://www.americanchemistry.com/s_acc/sec_article.asp?CID=985&DID=5343).

BBC h2g2. "Sir Humphry Davy." Retrieved November 20, 2008 (http://www.bbc.co.uk/dna/h2g2/A25568445).

BoneandSpine.com. "Calcium Hydroxyapatite Disease." Retrieved November 21, 2008 (http://boneandspine.com/arthritis/calcium-deposition-disease).

Canadian Lime Institute. "History of Lime." Retrieved September 20, 2008 (http://www.canadianlimeinstitute.ca/history.shtml).

Combs, Marianne. "Radium Girls Remembered." Minnesota Public Radio. Retrieved September 29, 2008 (http://minnesota.publicradio.org/display/web/2008/05/05/radiumgirlsremembered/?refid=0).

Curie, Marie. "The Discovery of Radium." Speech at Vassar College, May 14, 1921. *Internet Modern History Sourcebook*. Retrieved Nov. 12, 2008 (http://www.fordham.edu/halsall/mod/curie-radium.html).

DeCelles, Paul. "Chemical Bonds." The Virtually Biology Course. Retrieved November 5, 2008 (http://staff.jccc.net/PDECELL/chemistry/bonds.html#hydrobond).

Denver Museum of Nature & Science. "Largest Aquamarine Ever Found in North America Donated to Museum." Retrieved November 11,

2008 (http://www.dmns.org/main/en/General/Exhibitions/content/aquamarine.htm).

Emporia State University. *GO 340 Gemstones & Gemology: Beryl.* Retrieved September 30, 2008 (http://www.emporia.edu/earthsci/amber/go340/beryl.htm).

Jessey, Dave, and Don Tarman. "Mineral Identification." Geology Department of Cal Poly-Pomona. Retrieved November 23, 2008 (http://geology.csupomona.edu/alert/mineral/minerals.htm).

Jones and Bartlett Publishers. "Botany: An Introduction to Plant Biology: Glossary." 2008. Retrieved November 16, 2008 (http://biology.jbpub.com/botany/interactive_glossary_showterm.cfm?term=reducing%20power).

Los Alamos National Laboratory. "Periodic Table of Elements." Retrieved September 4, 2008 (http://periodic.lanl.gov).

MayoClinic.com. "Barium Sulfate." Retrieved December 2, 2008 (http://www.mayoclinic.com/health/drug-information/DR600237).

National Institutes of Health Office of Dietary Supplements. "Dietary Supplement Fact Sheets." Retrieved November 24, 2008 (http://ods.od.nih.gov/factsheets).

PBS. "A Science Odyssey." Retrieved November 12, 2008 (http://www.pbs.org/wgbh/aso/tryit/atom).

Royal Society of Chemistry. "Some Basic Chemistry." Retrieved September 2, 2008 (http://www.rsc.org/education/teachers/learnnet/cfb/basicchemistry.htm).

Royal Society of Chemistry. "A Visual Interpretation of the Table of Elements." Retrieved September 2, 2008 (http://www.rsc.org/chemsoc/visualelements/pages/pertable_j.htm).

Smithsonian Education. "Ocean Market." Retrieved November 15, 2008 (http://smithsonianeducation.org/educators/lesson_plans/ocean/market/proced.html).

Swiecki, Rafal. "Alluvial Exploration & Mining." Retrieved November 11, 2008 (http://www.minelinks.com/alluvial/aquamarine1.html).

Thomson Gale. "Chemical Elements: From Carbon to Krypton." 2005. Retrieved November 7, 2008 (http://www.chemistryexplained.com/elements/index.html).

U.S. Department of Energy. "Newton's Ask a Scientist." Retrieved November 5, 2008 (http://www.newton.dep.anl.gov).

U.S. Environmental Protection Agency. "A Citizen's Guide to Radon." Retrieved November 17, 2008 (http://www.epa.gov/radon/pubs/citguide.html#howdoes).

U.S. Environmental Protection Agency. "Ottawa Radiation." Retrieved November 17, 2008 (http://www.epa.gov/superfund/accomp/factsheets05/ottawa.htm).

U.S. Environmental Protection Agency. "Radiation Protection." Retrieved November 17, 2008 (http://www.epa.gov/rpdweb00/understand/pathways.html).

U.S. Environmental Protection Agency. "Strontium." Retrieved November 17, 2008 (http://www.epa.gov/rpdweb00/radionuclides/strontium.html#wheredoes).

Western Oregon University. "A Brief History of the Development of the Periodic Table." Retrieved September 15, 2008 (http://www.wou.edu/las/physci/ch412/perhist.htm).

Winter, Mark. "Web Elements: The Periodic Table on the Web." The University of Sheffield and Web Elements Ltd., 1993–2008. Retrieved September 2, 2008 (http://www.webelements.com).

Zb.Zwolinski. "Maria Sklodowska-Curie, 1867–1934." Retrieved November 12, 2008 (http://www.staff.amu.edu.pl/~zbzw/ph/sci/msc.htm).

Index

A

alkaline earth metals
  as compounds, 5, 7, 13, 17–18, 19,
    26–31, 32
  makeup of, 14–16, 17, 18
  in periodic table, 5, 6, 16
  properties of, 11–13, 18–20
  reactivity of, 6, 8, 13, 14, 16, 19, 20
  as reducers, 13, 19–20, 23

B

barium
  compounds of, 30, 36–37
  electrons in atom of, 14
  isolation of, 9, 24, 25
  properties of, 19, 20
  uses for, 9, 24, 30, 36, 37
  where it's found, 20, 24, 37
beryllium, 13
  compounds of, 27–28
  electrons in atom of, 14
  isolation of, 21
  properties of, 18, 19–20, 21
  uses for, 21
  where it's found, 6, 32

C

calcium, 13, 16, 23–24, 26
  compounds of, 29, 34
  electrons in atom of, 14

  importance of for humans, 34, 35
  isolation of, 9, 23, 25
  properties of, 19, 20
  uses for, 23, 29
  where it's found, 8–9, 11, 19, 23, 34, 35
Curie, Marie, 9, 10, 18, 24–25

D

Davy, Sir Humphry, 7, 9, 25

E

Epsom salt, 6, 7, 28, 29, 33

F

fireworks, 6, 9, 22, 30, 36, 37

I

ionic bonds, 17–18, 19, 22, 27

L

lime/limestone, 8–9, 19, 23, 29, 34

M

magnesium, 13, 18
  compounds of, 28, 29, 33–34
  electrons in atom of, 14
  isolation of, 13, 22, 23, 25
  properties of, 19, 20
  uses for, 13, 22, 28
  where it's found, 7, 8, 11, 19, 22,
    33–34, 35
Mendeleyev, Dmitry, 4, 11

R

radium, 18
   compounds of, 30–31
   dangers of, 10, 18, 19, 20, 25, 31, 37
   discovery of, 9
   electrons in atom of, 16
   isolation of, 24
   properties of, 19, 20, 25
   uses for, 10, 25, 30–31
   where it's found, 37

S

strontium
   compounds of, 29–30, 36
   dangers of, 23–24, 36
   discovery of, 9, 24
   electrons in atom of, 14
   isolation of, 23, 25
   properties of, 19
   uses for, 23, 24, 29–30
   where it's found, 9, 23–24, 29, 36

About the Author

Bridget Heos became interested in chemistry when her high school teacher Sister Harriet stuck her hand in a beaker of boiling liquid and didn't get burned. She was demonstrating the unique characteristics of compounds, in this case the low boiling point. For Heos, a career in nonfiction writing—first as a newspaper reporter and then as a children's book writer—has allowed her to revisit chemistry and other science disciplines.

Photo Credits

Cover, pp. 1, 12, 15, 38–39 by Tahara Anderson; p. 5 Hulton Archive/Getty Images; p. 7 © Charles D. Winters/Photo Researchers, Inc.; pp. 8, 18, 22, 27, 28 Wikimedia Commons; p. 10 © Bettmann/Corbis; p. 17 © Andrew Lambert Photography/Photo Researchers, Inc.; p. 20 © sciencephotos/Alamy; p. 23 Chuck Nacke/Time & Life Pictures/Getty Images; p. 31 Wikipedia; p. 33 © Scott Dressel Martin/Bailey Archive, Denver Museum of Nature & Science; p. 34 © www.istockphoto.com/Sean Locke; p. 36 © Medical Body Scans/Photo Researchers, Inc.

Designer: Tahara Anderson; Editor: Bethany Bryan;
Photo Researcher: Amy Feinberg